F-102 Delta Dagger

Written by Ken Neubeck

Cover Art by Don Greer
Profile Illustrations by Tom Tullis
Line Illustrations by Matheu Spraggins

(Front Cover) An F-102 Delta Dagger from the 317th FIS based in Elmendorf AFB in Alaska flies over the arctic region to intercept Soviet aircraft.

(Back Cover) An F-102 fires an air-to-air Falcon missile over Vietnam. Aircraft in Vietnam were painted in the Southeast Asia (SEA) camouflage paint scheme during the conflict.

About the Walk Around®/On Deck Series®

The Walk Around®/On Deck® series is about the details of specific military equipment using color and black-and-white archival photographs and photographs of in-service, preserved, and restored equipment. *Walk Around®* titles are devoted to aircraft and military vehicles, while *On Deck®* titles are devoted to warships. They are picture books of 80 pages, focusing on operational equipment, not one-off or experimental subjects.

Copyright 2010 Squadron/Signal Publications
1115 Crowley Drive, Carrollton, TX 75006-1312 U.S.A.
Printed in the U.S.A.

All rights reserved. No part of this publication may be reproduced, stored in a retrieval system, or transmitted in any form by means electrical, mechanical, or otherwise, without written permission of the publisher.

ISBN 978-0-89747-615-7

Military/Combat Photographs and Snapshots

If you have any photos of aircraft, armor, soldiers, or ships of any nation, particularly wartime snapshots, please share them with us and help make Squadron/Signal's books all the more interesting and complete in the future. Any photograph sent to us will be copied and returned. Electronic images are preferred. The donor will be fully credited for any photos used. Please send them to:

Squadron/Signal Publications
1115 Crowley Drive
Carrollton, TX 75006-1312 U.S.A.
www.SquadronSignalPublications.com

(Title Page) F-102A aircraft is stationed at Don Muang Air Base Thailand and is painted in the SEA camouflage scheme. (National Archives via Dennis R. Jenkins)

Acknowledgments

I wish to thank Major George Worrall and the Connecticut ANG unit at Bradley along with Captain Cory Angell and the Pennsylvania National Guard at Indiantown Gap for allowing me access to photograph their F-102A exhibits. Thanks go to Michael Benolkin, John A. Gourley III, Dennis R. Jenkins, David Kidman, and Ben Peck for the various photos that they have provided. Additional archival photographs were made available from the National Museum of the Air Force and the National Archives.

Introduction

As one of the U.S. Air Force Century Aircraft series, the Convair F-102A Delta Dagger was developed as an interceptor able to counter the threat of long-range Soviet bombers during the Cold War. It was also the first U.S. production aircraft to feature the delta wing design.

Following the Soviet Union's first nuclear test in 1949, the U.S. Air Force (USAF) initiated a crash program to confront possible strategic threats from the Soviets. In high demand was an aircraft that could intercept Soviet long-range nuclear-capable bombers. This interceptor had to be capable of speeds exceeding Mach 1 and its ceiling had to exceed 50,000 feet. Convair was one of several companies that submitted proposals in response to the Air Force request for such an interceptor to be ready by 1954.

Convair based its design on the earlier XP-92A experimental aircraft that featured a triangular or delta wing swept back at 50 degrees. That feature reduced drag coefficient and produced greater stability at very high altitudes. Convair won the contract, building and flying six YF-102 prototypes by 1953. These prototypes failed to break the sound barrier during level flight, however, and speeding them up required several design changes. Beside redesigning the air intake and the canopy, company engineers modified the shape of the YF-102A aft fuselage by adding two bulges (called "love handles") that would use the area rule of aerodynamics to help the aircraft exceed Mach 1. The plane finally entered production as the F-102A.

The F-102A was powered by a single Pratt & Whitney J57-P-23A engine, internally carried four AIM-4 missiles for intercept missions, and had a crew of one – the pilot.

As the delta-wing F-102A was radically different from other aircraft in service at the time, the Air Training Command saw a need for a trainer version of the F-102A. Responding to the need, Convair developed the two-seat TF-102A. Its growing pains were similar to those experienced by the original F-102 in terms of directional control and stability. The fuselage had to be redesigned to accommodate side-by-side seating and other modifications had to be made in order to correct buffeting problems at high speeds.

The first TF-102A prototype took to the air in June of 1955, paving the way for a total of 111 TF-102A aircraft that would eventually be produced, out of the total number of 899 Delta Daggers. In September of 1956, Colonel Charles Rigney set a record in an F-102, flying between George AFB, California, and Oklahoma City, at an average speed of 820 mph.

Because of its role as an interceptor, the F-102A deployed in many cold-weather locations such as Iceland, Greenland, and Alaska. The F-102A also saw service in warmer climes. The USAF supplied TF-102As to NATO Allies Turkey, in 1968, and Greece, in 1969. Over the course of the Cold War, the aircraft would intercept Soviet bombers on thousands of occasions. Initially sent to Southeast Asia to protect bases from enemy aircraft during the Indochina War, the plane later took part in ground attacks on buildings and other installations in Vietnam.

Beginning in 1960, F-102A aircraft would be reassigned to the Air National Guard (ANG) throughout the United States. Prior to retirement in 1976, the aircraft served with 22 ANG squadrons. Then, in response to the need for a high-speed drone in 1978, the F-102A was converted to this new role under the Pave Deuce program. F-102As were stripped of armaments and electronics and equipped for remote control by a pilot on the ground. A total of 152 aircraft, designated PQM-102A and PQM-102B, were converted to drones.

Two F-102A aircraft from the 57th Fighter Interceptor Squadron take off from a U.S. airfield in the early 1960s. (National Archives via Dennis R. Jenkins)

This F-102A serves with the 317th Fighter Interceptor Squadron based in Elmendorf AFB in Alaska in January 1958. Many of the interceptors were assigned duty in cold-weather locations, ready to counter possible Soviet threats. (USAF)

Air National Guard (ANG) Pilot James Brundige was the last pilot to fly this F-102A (serial number 57-788), here sitting at the front gate at the New York ANG base in Westhampton Beach. While in the pattern for landing, there was an electrical fire that progressed into a hydraulic fire. Although the pilot managed to land the aircraft, the internal damage was too great, and the plane was relegated to the role of static display at the front gate in the late 1970s. The aircraft has gone different several paint schemes over the years since then, with the current scheme featuring the NY ANG emblem on the tail. (Ken Neubeck)

F-102A Delta Dagger

Wingspan:	38 feet, 1.6 inches
Length	68 feet 1.8 inches
Height	18 feet, 10.5 inches
Empty weight	20,234 pounds (with drop tanks)
Combat weight	31,276 pounds (with drop tanks)
Powerplant:	One Pratt & Whitney J57-P-23A turbojet
Armament:	Twelve 2.75-in folding fin aircraft rockets and Two AIM-26A.b Super Falcon or One AIM-26A/b + two AIM-4A/B radar-guided Falcon missiles or Three AIM –4A + three AIM-4C/D/G infra-red Falcon missiles
PERFORMANCE	
Maximum Speed:	825 MPH (Mach 1.3)
Service Ceiling:	53,400 feet (9390 M)
Range:	1,492 miles
Crew:	One
Number built	889

TF-102 Delta Dagger

Wingspan:	38 feet, 1.6 inches
Length	63 feet 4.5 inches
Height	21 feet, 2.5 inches
Empty weight	21,062 pounds (with drop tanks)
Combat weight	32,104 pounds (with drop tanks)
Powerplant:	One Pratt & Whitney J57-P-23A turbojet
Armament:	Twelve 2.75-in folding fin aircraft rockets and Two AIM-26A.b Super Falcon or One AIM-26A/b + two AIM-4A/B radar-guided Falcon missiles or Three AIM –4A + three AIM-4C/D/G infra-red Falcon missiles
PERFORMANCE	
Maximum Speed:	770 MPH (Mach 1.0)
Service Ceiling:	51,400 feet (9390 M)
Range:	1,466 miles
Crew:	Two
Number built	111

At the front of the F-102A is a black radome cover, attached to which is a long pitot static tube. (Ken Neubeck)

The radome, made of fiberglass, is seen together with the pitot tube in this close up shot of the front of the aircraft. (Ken Neubeck)

A series of screws, rather than quick-release latches, secures the fiberglass radome to the forward fuselage of the aircraft. A number of electronic components are situated on the frame underneath the radome, including the automatic flight control system (AFCS) and the automatic instrument landing approach system (AILAS), along with some components of the fire control system. (Ken Neubeck)

The long pitot static boom assembly on the F-102A's radome features red and white stripes in a "barber pole" pattern. The pitot tube measures air pressure, enabling the pitot static system to determine air speed and altitude of the aircraft while in flight. (Ken Neubeck)

The basic paint scheme for the front probe is a barbershop stripe pattern of red and white. An attachment fitting connects the probe to the front cover. (Ken Neubeck)

One ground crew member attaches an engine inlet insert cover while the other slips a protective cover over the probe. (National Archives via Dennis R. Jenkins)

7

This F-102A Delta Dagger, Serial Number 59-984, is on display at the Wings Over the Rockies Air and Space Museum in Denver, Colorado, and represents one of the better-condition F-102A aircraft that can still currently be seen. The inside of the rear canopy section is visible here. Falcon missiles are loaded in the left missile bay, and some 2.75- inch rockets are present in the left folding fin panel. (Ben Peck)

This head-on shot of the F-102A shows that the air inlet scoops are located on the sides of the aircraft and blended with the fuselage. (Ken Neubeck)

Early-production F-102As suffered from a vibration condition that caused a fuselage buzz. This was fixed by adding a splitter plate, extending from the air inlet scoop. (Ken Neubeck)

The air inlet scoops on the TF-102A trainer are located on the lower part of the fuselage and protrude away from the body of the aircraft, in contrast to the air inlet configuration on the earlier F-102A seen at left. (David Kidman)

The TF-102A inlet scoops were moved lower and an indentation was built into the fuselage to eliminate the splitter plate needed on the earlier F-102A. (David Kidman)

The front of the canopy for the single-seat F-102A aircraft is narrow and pointed, with two flat glass panels for both the right and left canopy sections. (Ben Peck)

Located on the inside of each front glass panel is a series of metal strips used for de-icing and defogging the windshield. (Ben Peck)

Located just below the top of the canopy is the optical sight assembly that is used by the pilot to spot targets. (Ken Neubeck)

The top of the pilot's seat is visible through the glass in the rear of the canopy. The canopy sections are fitted using standard hardware around the frame. (Ken Katz)

Located below the right side of the canopy of the F-102A is a quick release access door for the manual canopy unlatch handle. (Ken Neubeck)

Ground crew can rescue a pilot trapped inside a cockpit by accessing this emergency panel located above the left wing. (Ken Neubeck)

On the TF-102A, the emergency ejection seat activation is located inside a quick release panel on the left side of the canopy, in front of the warning triangle. (Ken Neubeck)

Located on the left side of the two-piece rear canopy section is an emergency canopy jettison handle marked in yellow and black. (Ben Peck)

The front fuselage section of the F-102A aircraft was designed to be long in order to accommodate access doors and the splitter plate in front of the engine intake. (Ken Neubeck)

The front fuselage section on the TF-102A is significantly shorter than that on the F-102A. The splitter plate in front of the intake is also dispensed with on the TF-102A. (Ken Neubeck)

This view of the right side of the forward fuselage of the F-102A aircraft shows a long and a short access door, along with the aircraft's long nose section to accommodate fire control equipment. (Ken Neubeck)

This right side view of the TF-102A aircraft shows that the shorter access door found in the F-102A aircraft has been eliminated from the design of the shorter nose section of the two-seater TF-102A. (Ken Neubeck)

When the F-102A design was modified into the TF-102A trainer, the front section was widened to accommodate side-by-side seating in the cockpit. (Ken Neubeck)

The widened TF-102A front fuselage, redesigned engine intakes, and indented fuselage structure are apparent in this view from underneath. (Ken Neubeck)

The forward windshield section of the TF-102A canopy features a two-piece center section flanked by a left and right triangular window. (Ken Neubeck)

The rear canopy section of the TF-102A is also wider and rounder than the corresponding canopy section of the original, single-seat F-102A. (Ken Neubeck)

An early production F-102A, serial number 53-1793, has just launched a missile. The aircraft was taking part in missile firing tests in 1955. (USAF)

The missile is now being propelled forward of the aircraft. This particular aircraft was destined to be lost in a crash in New Mexico in December of 1955. (USAF)

Located in equipment bays on both the right and left side of the forward fuselage of the F-102A aircraft is the Hughes-design MG-3 fire control equipment for the missile launching from the F-102A. (Ken Neubeck)

The fire control allows for the automatic door opening and extension of the missile launchers for firing AIM-4 Falcon missiles. In the foreground is a Super Falcon AIM-26 missile, which could carry a nuclear warhead and substitute for the AIM-4. (Ben Peck)

A splitter plate installed on the F-102A to solve fuselage vibration issues created a new problem - the splitter plate overlapped the access doors on both sides of the aircraft. Above is the right side. (Ken Neubeck)

The unique arrangement of the front portion of the splitter plate was on both sides of the aircraft, and the attachment points for this plate can be seen in this left side view. (Ken Katz)

The armament access door on the left side is covered by the splitter plate which is divided into two sections, with the shiny side actually being attached to the access door by two fasteners. (Ken Neubeck)

The TF-102A's design by Convair engineers have unique contours and eliminate the need for the splitter plate. The structure reduced vibration. The intake area of the engine cooling inlets only required a stainless steel rim. (Ken Neubeck)

Each of the engine inlet scoops on the F-102A had a splitter plate that extended forward from the inboard side of each scoop. This plate was added during F-102A production in response to aerodynamic problems with the fuselage. The inboard side of the inlet was slightly curved, following the contour of the fuselage. Two attachment points at the lower and upper part of the inlet connect the plate to the fuselage. (Ken Neubeck)

By contrast, the engine inlet scoops on the TF-102A aircraft do not have a splitter plate on the inboard side of the inlet. This is because both the fuselage area in front of the inlets and the inlets themselves were designed with aerodynamic curves to allow for better airflow intake. The TF-102A inlet is narrower than the inlet on the F-102A model. Also, a stainless steel ring lines the entire inlet area. (Ken Neubeck)

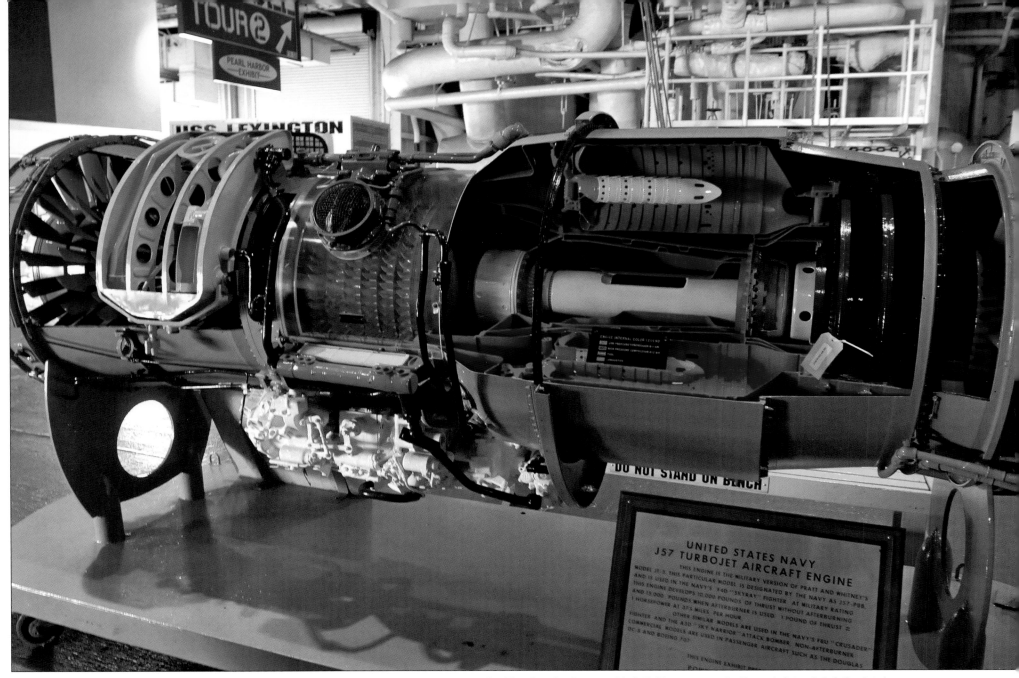

A cutaway of a Pratt & Whitley J-57 engine displayed at the USS Lexington Museum On The Bay in Corpus Christi, Texas, reveals (from left to right) the intake, compressor, diffuser, and burner sections. The turbine and afterburner sections were aft of the main engine. The J-57 engine is an axial flow gas turbine, commonly known as a two-spool engine, since it has two rotors revolving on concentric shafts. (Mark M. Hancock)

The sixth YF-102 aircraft conducts a flight test over the desert near Edwards AFB, California. This aircraft has an all-metal finish with the Convair company markings. (National Archives via John A. Gourley III)

This later YF-102A is retracting its landing gears after takeoff. The design of the canopy tail has changed from those on the earlier, pre-production aircraft. (National Archives via John A. Gourley III)

This YF-102A aircraft is being prepared for takeoff from Edwards AFB, California, in about 1959. The front canopy design, which has cross sections that are parallel to the center frame and would be changed in later pre-production aircraft. The rear fuselage is straight on this aircraft and this would be modified significantly during the YF-102A test program aircraft to deal with aerodynamic issues concerning the delta wing design and the rear fuselage. (National Archives via John A. Gourley III)

The boarding ladder used on the TF-102A is the same as that used on the F-102A with two being used that go around the inlet scoops as seen above. (USAF)

In this photo of the TF-102A and the F-102A, it can be seen that the front of the TF-102A in the foreground is dramatically wider and shorter than the front of the F-102A in the background. With a shorter nose section, some of the avionic equipment for the TF-102A had to be relegated to the area behind the cockpit. (National Archives via Dennis R. Jenkins)

As this F-102A aircraft takes off, the three landing gears are retracted into the aircraft. The two main landing gears, with wing panels attached to each strut, retract inboard into the wheel well. A pair of doors hinged in the middle of the fuselage close during the main landing gear retraction. In this shot, the nose landing gear has receded into its wheel well and the door is in the process of closing. (USAF)

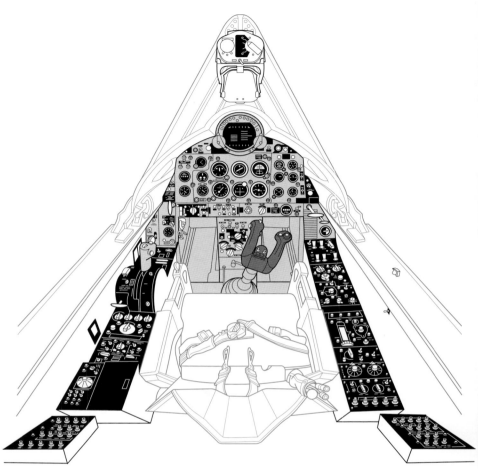

The F-102A cockpit is very narrow, as can be seen by the way that the seat fits underneath the narrow canopy. (Ben Peck)

This drawing above and the USAF photo at right illustrate the layout of the F-102A cockpit and the locations of some of the most important instruments on the Delta Dagger. The control stick consists of a dual handle control that requires both hands and is located between the pilot's legs. Located in front of the dual handle control are the utility panel and the lighting control panel. The throttle controls are located at the left hand of the pilot. The optical sight is high above the console, just above the radar scope. Standard cockpit instruments, including the airspeed-Mach indicator, attitude indicator, fuel quantity gage, hydraulic pressure gage, and altimeter, are located on the main panel. Circuit breaker panels for different functions are located behind the pilot's seat on both the left and right sides of the rear wall. There is yet another circuit breaker panel located in the left-side leg area.

The trainer version TF-102 aircraft features an enlarged cockpit in order that accommodates side-by-side seating for the pilot instructor and student. (Ken Neubeck)

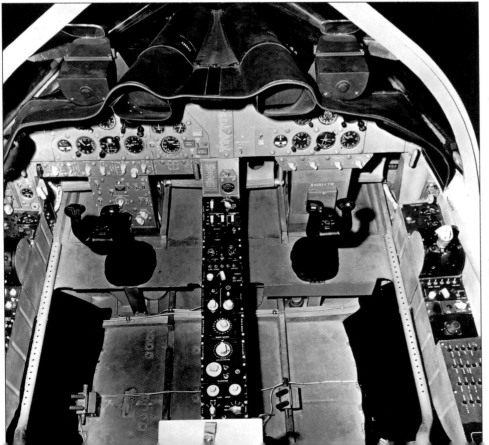

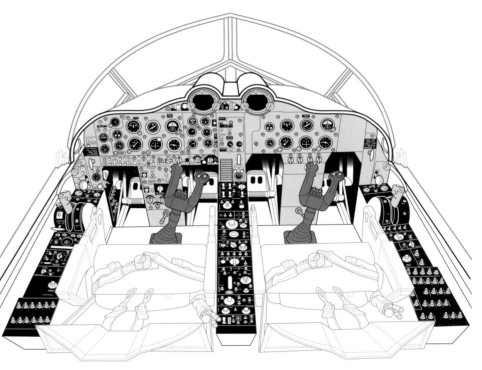

As the TF-102A aircraft is a side-by-side seating trainer, common flight instruments shared by both crewmembers are located on the center part of the front instrument panel as well as the center console. In addition, there are two sets of control sticks as well as two sets of rudder pedals for each of the crewmembers. The pilot instructor sits on the left side of the aircraft while the student sits on the right. There is a throttle quadrant located on the right hand side for the student whereas the pilot instructor's throttle control remains on the left hand side of the pilot. Instead of one radar scope, each crewmember has his own small individual radar scope located on the top of the main instrument panel. There are two individual optical sights on each side as well. The pilot instructor has a utility panel located in front of his control stick; there is no utility panel for the student. The seats have been removed in the photo at left. (U.S. Air Force)

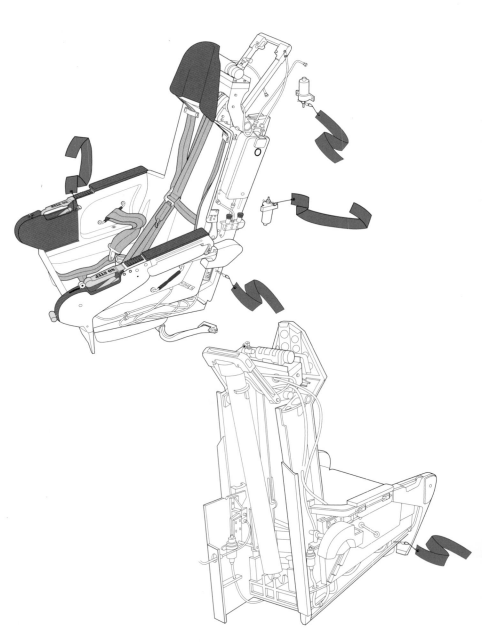

On this chart of major components of the ejection seat designed by Weber Aircraft and used in the F-102A, the red flags indicate the safety pins installed when the aircraft is on the ground and removed prior to flight. The ejection seat could break through the canopy if it failed to jettison properly during the ejection sequence.

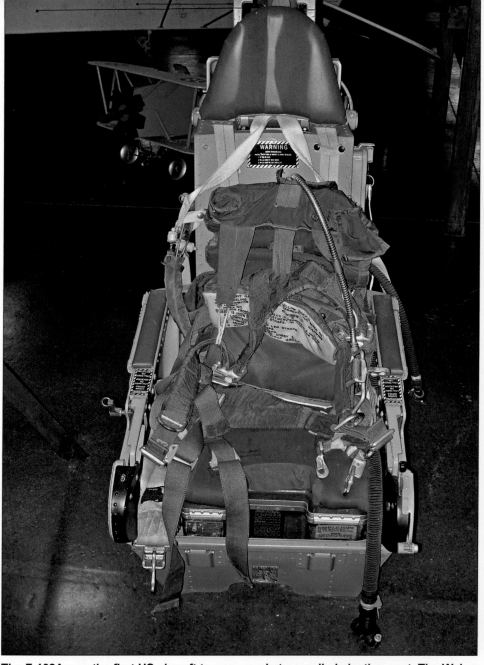

The F-102A was the first US aircraft to use a rocket-propelled ejection seat. The Weber Aircraft-manufactured seat consisted of a gray colored seat bucket with red head rest and red arm rests. Restraining straps are in olive green and the seat cushion is light brown. (Michael Benolkin)

This aircraft, serial number 56-1277, is from the 317th Fighter-Interceptor Squadron (FIS) based at Elmendorf AFB in Alaska in January 1964. The 317th FIS was transferred from Washington state to Alaska in 1957 when it received F-102 aircraft. The 317th FIS was deactivated in 1969. (National Archives via John A. Gourley III)

The first flight of the YF-102 (52-7994) takes place at Edwards AFB in October 1953. (National archives via Dennis R. Jenkins)

An early F-102 model undergoes wind tunnel testing at Langley AFB. Tests revealed that the original design would not reach Mach 1 speed and that modifications to the fuselage would be required in production. (National Archives via Dennis R. Jenkins)

An F-102A from the 57th FIS at Keflavik, Iceland, fulfills its primary role of interceptor as it shadows a Soviet Air Force TU-95 long-range bomber during the 1960s. The F-102A performed thousands of intercepts like this. (U.S. Air Force)

Personnel crowd around F-102A, serial number 54-379, after its landing at George AFB, northeast of Los Angeles, California. (National Archives via Dennis R. Jenkins)

These three F-102A aircraft from the 332nd FIS flew till 1965 when all three aircraft were converted to PQM-102A target drones. (National Archives via Dennis R. Jenkins)

These F-102As serve the last active F-102A unit, the 526th Fighter Interceptor Squadron, based at Ramstein AFB, West Germany. (National Archives via Dennis R. Jenkins)

Ground personnel survey damage on an F-102A after its right main landing gear collapsed during testing. (National Archives via Dennis R. Jenkins)

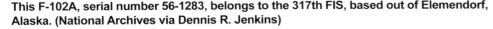

Five F-102A aircraft from different squadrons await their next missions on the airfield of George AFB, not far from Los Angeles, in California. (National Archives via Dennis R. Jenkins)

This aircraft, serial number 54-1394, would serve with the Idaho Air National Guard, one of many ANG units in which the F-102 served during the 1960s. (National Archives via Dennis R. Jenkins)

This F-102A, serial number 56-1283, belongs to the 317th FIS, based out of Elemendorf, Alaska. (National Archives via Dennis R. Jenkins)

Two F-102As maneuver in the Arctic Zone, a common destination for the aircraft playing the interceptor role in the Cold War. (National Archives via Dennis R. Jenkins)

The rear part of the F-102A canopy is a hinged hatch that pivots when opened. Inside the door under the right side of the canopy is the manual canopy release. (Ken Neubeck)

This is the canopy hinge area where the canopy pivots when opened. Three vents in the rear portion of the canopy aid in cooling. (Ken Neubeck)

Behind the hinge area of the F-102A is an access panel for ground maintenance access to the upper electronic equipment bay. (Ken Neubeck)

To improve air flow and accommodate extra equipment inside the upper fuselage, the area behind the canopy of the TF-102A is larger than that on the F-102A. (Ken Neubeck)

Four F-102A Delta Daggers from the 4780th Air Defense Wing (ADW) fly a training mission over downtown Dallas, Texas, in 1965. This wing was based out of Perrin AFB in Texas, where the primary interceptor school was located. The tail markings for this wing are very distinctive with a white band and a red arrow and a green arrow over the band. (National Archives via Dennis R. Jenkins)

Three tanks in the delta wings of the F-102A hold over 1,000 gallons of fuel. (National Archives via Dennis R. Jenkins)

The TF-102A trainer appears here on the left and the F-102A aircraft is on the right. The fuselage for the TF-102A is wider and shorter than that of the F-102A, but the wing design remains the same. (National Archives via Dennis R. Jenkins)

This overhead view of a production F-102A aircraft undergoing ground maintenance in the late 1950s clearly shows the delta wing design of the aircraft built around a fuselage with a shape reminiscent of a Coca-Cola bottle. Also visible are all four fence structures on the wing. The elevon surfaces at the back of both wings are extended downward. The typical wing markings of the Delta Dagger in U.S. service for this time period were USAF on the top of the right wing and the Air Force national insignia on the top of the left wing. (National Archives via Dennis R. Jenkins)

Tasked to defend the United States from attacks over the North Pole, the F-102A deploys in the frigid Arctic regions from Alaska to Greenland during the 1950s, 1960s, and 1970s. The aircraft and its systems were designed to endure those areas' extreme temperatures, which often dropped many degrees below freezing. (National Archives via Dennis R. Jenkins)

This YF-102 is one of the two prototypes that Convair built at the start of the program in the 1950s. The consistently smooth fuselage proved aerodynamically problematic.

Convair personnel fit a probe behind the canopy of an F-102A test aircraft to give it in-flight refueling capability. (National Archives via Dennis R. Jenkins)

A KC-135 tanker refuels a production F-102A. In-flight refueling premiered late in the program on some F-102As deployed abroad. (National Archives via Dennis R. Jenkins)

These three F-102A aircraft belong to the Iceland-based 59th FIS. (National Archives via Dennis R. Jenkins)

F-102A, serial number 53-1793, of the 509th FIS sits on the tarmac of Don Muang Royal Thai Air Force Base just north of Bangkok, Thailand, during the Vietnam War. F-102A aircraft were sent here to bolster Thai air defenses as early as 1961. This aircraft is painted in the Southeast Asia (SEA) camouflage paint scheme and is equipped with external fuel tanks on the pylon of each wing. The aircraft is fitted with an additional position light on the top portion of the forward fuselage. The air brake is deployed in the open position and access doors located on the lower fuselage are open as well. (National Archives via Dennis R. Jenkins)

This TF-102A has crew boarding ladders, one on the left and one on the right sides of the aircraft, to accommodate the two pilots. (USAF)

The drogue parachute billows out from the open air brake panels as this first-production TF-102A aircraft, serial number 54-1351, comes in for a landing. (USAF)

A total of 111 TF-102As were built to facilitate training in the 788 F-102A aircraft. A number of them are on display throughout the United States. This TF-102A is part of the Century Circle display at the Edwards AFB museum in California. (David Kidman)

A TF-102A (foreground) is parked next to an F-106B Delta Dart trainer at the Edwards AFB museum. In the 1960s and 1970s, the Delta Dart took over the role played by its predecessor, the F-102A. (David Kidman)

This forward view of the nose landing gear assembly for the F-102A Delta Dagger illustrates how the wheel assembly is side-mounted to the strut that curves around the wheel. The single door for the nose landing gear is located on the left underside of the aircraft. The forward taxi light is attached to the nose landing gear door by a bracket assembly. In this front view of the aircraft, some of the blue 2.75-inch rockets can be seen loaded and protruding from the folding fin panels hanging from the aircraft. Two more rockets are situated on floor display stands. (Ben Peck)

The early F-102A nose wheel is a two-piece assembly with six cutouts. (Ken Neubeck)

Later F-102A nose wheel designs are two-piece wheel assemblies with machined spokes as part of each wheel half. (Ben Peck)

The nose landing gear consists of a single fork that holds the wheel from the left side. This fork curves around the top of the wheel and fits inside of the nose landing gear strut assembly. When the landing gear is deployed on the ground, there is about six inches of exposed shiny metal surface on the fork section, which pulls in during retraction of the landing gear when the aircraft is in flight. (Ben Peck)

The scissors assembly is connected to the top of the fork section and to the bottom of the strut assembly. The scissors drive the lower portion of the landing gear away from the strut portion when the gear is fully extended. Located in the rear of the strut is the nose wheel steering unit, a cylinder in metallic finish, mounted on the upper portion of the strut by means of a dual bracket assembly. (Ben Peck)

In the lower part of the strut area, above the ring and bearing, is a gear with teeth that engages with the nose wheel steering mechanism. There is a tie down ring that is located just above this gear. (Ben Peck)

The nose landing gear panel door covers the nose landing gear from the left and extends towards the front of the aircraft. This view from the right of the aircraft shows the position of the nose gear, door assembly, and the taxi light located on the door. (Ken Katz)

The taxi light for the nose gear is mounted inside a protective bracket on the middle of the nose landing gear door. (Ken Neubeck)

Another wheel design for the nose landing gear, with more spokes and no open slots, was used in later-model F-102A and TF-102A aircraft. (David Kidman)

The upper portion of the strut assembly goes into the wheel well. Connected to the front of the strut assembly is a two-piece linkage assembly that pulls the gear into the well during retraction. The top of the strut is a T-section that pivots inside the wheel well. (David Kidman)

A nut assembly with a cotter pin connects the two links in the nose landing gear assembly. The top link in the assembly pivots on a bracket inside of the wheel well. Hydraulic lines are carefully routed and secured to the links and the top of the strut assembly. (Ken Neubeck)

A triangle bracket on the door panel connects to a curve link located on the left side of the aircraft. The curve link pulls the door closed during gear retraction. (David Kidman)

The linkages connecting the nose landing gear strut are all mounted inside of the wheel well area. (Ken Neubeck)

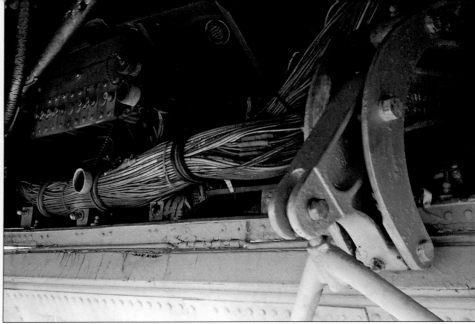

Located on the left side of the aircraft, just above the nose gear door and inside the wheel well, are electrical harnessing and a circuit breaker box. (Ken Neubeck)

There is room to accommodate a number of electronic boxes associated with the landing gear systems such as signal generators that are tucked away on the left side and towards the front of the deep, roomy wheel well. (Ken Neubeck)

Many of the links and hydraulic components that are used to open and close the nose landing gear are located near the the ceiling of the nose gear wheel well. The white two-piece link assembly connects the strut to a roller assembly at the ceiling and is driven by a hydraulic actuator. The top of the strut itself is a T section that rotates in place at the top of the wheel well and is driven by links located on each side of the strut. (Ben Peck)

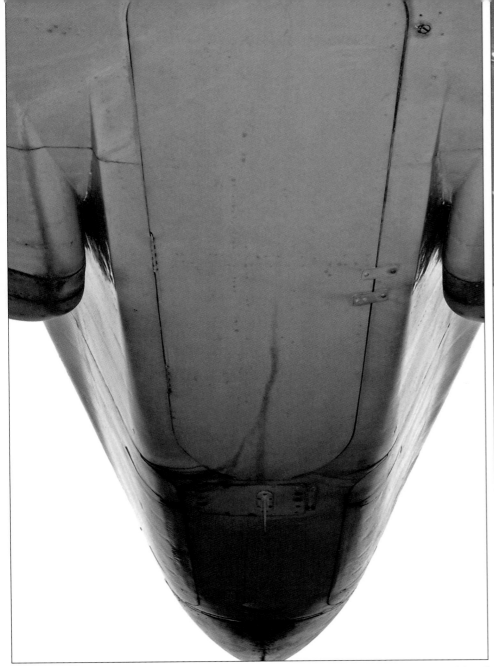 

The entire nose landing gear door in the closed position is seen on the underside of a TF-102A. Both engine air inlet cooling scoops begin by the midpoint of the landing gear door. The different fuselage surface panels that make up the lower part of the inlet scoop areas are contoured to meet the shape of the inlet scoops. (Ken Neubeck)

In this view of the lower fuselage, looking from the rear exhaust section towards the front of the aircraft, it can be seen that the stainless steel exhaust section is mounted flush with the rest of the aircraft. In front of the exhaust section is the mounting base for two UHF antennas that have been removed from this aircraft. (Ken Neubeck)

From the side, it can be seen that the main landing gear assembly is slightly ahead of the inboard fence assembly that is located on the wing. (Ken Neubeck)

The uniquely shaped panel on the outside of the wheel assembly closes over the landing gear wheel well and meshes with the contours of the fuselage. (Ken Neubeck)

The door panel assembly for the main landing gear is hinged at the wing and is pulled closed towards the wheel well by linkages and actuators during flight. This particular display aircraft has a stabilizer bar that connects the two main landing gear wheels that are situated on jacks. The wheel bearing extends partially from the wheel and the wheel assembly is held together by 13 bolts. (Ken Neubeck)

As illustrated by this view of the right main landing gear assembly, both landing gears have scissor assemblies that are located on the inboard side of the gear and are partially extended when on the ground. The scissor assembly is retracted when the landing gear is pulled up during flight. The taxi light is located between the outboard panel and the main landing gear strut for both assemblies. (Ben Peck)

Compared to the nose landing gear wheel, the main landing gear wheel has a much larger bearing assembly that is located on the inboard side of the wheel and attaches to the bottom of the landing gear strut. Specified tire pressure for the two main wheel tires is 195 psi - higher than the pressure in the nose wheel tire. There are tie rings on the lower and upper portion of the inboard side of the landing gear. (Ben Peck)

The curved extrusion that connects the outboard door panel to the strut assembly of the landing gear is visible in this view from the side of the left main landing gear. Two linkage arms connected to the main landing gear strut pull the gear into the wheel well after takeoff. In the front of the gear inside the wheel well is an actuator that aids in moving the gear into the wheel well. (Ben Peck)

Scissors are extended about six inches when the aircraft is on the ground. The outside of the wheel bearing has a spring-loaded bumper to absorb shock when the gear is pulled up. (Ben Peck)

The landing gear strut is made by Menasco, a well-known manufacturer of struts. Servicing instructions are provided on a metal plate located on the strut. (Ben Peck)

Many of the hydraulic lines are located on the rear portion of the main landing gear with a flexible rubber line going from the base of the strut to the wheel assembly. Rigid metal lines are located in the middle of the strut. The curved extrusion that connects the door panel to the landing gear strut rotates on a pin that is connected to the top part of the scissors assembly. (Ben Peck)

Two linkages are connected to the rear of the strut, as seen in this view from the rear of the right main landing gear. Both linkages swivel towards the fuselage when the landing gear retracts during flight. The landing gear actuator is extended from the wheel well. The taxi light housing is attached to the inside of the outboard door panel assembly with the power wire for the light being routed on the inside of the door panel. (Ben Peck)

Close view of the scissors assembly shows the triangle linkage setup that is used to extend and retract the gear. Stainless steel clamps on the strut hold some of the lines in place. Located on the inboard side of the strut and behind one of the tie down rings is a service plug that is removed for hydraulic fluid replenishment during ground maintenance. (Ben Peck)

The construction of the main landing gear door panel assembly is rugged and consists of fairings and extrusions that are riveted together. About mid-way on the door is the linkage that is attaches the door to the main landing gear strut. There is about one inch of edging around the door that fits into the wheel well when the door is closed. (Ken Neubeck)

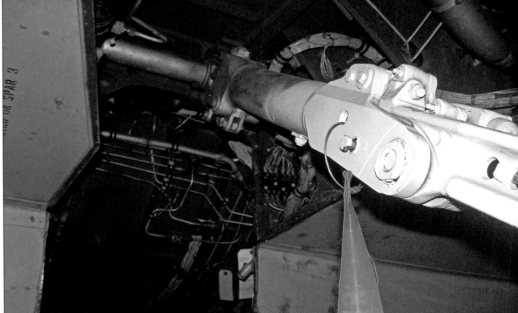

A pair of access doors hinged on the middle fuselage allow maintenance access to the lower part of the afterburner section of the engine, and to some of the hydraulic lines that are routed through this area. These access doors interface with each of the main landing gear door panels. The two-piece linkage connects inside the upper part of the wheel well in the fuselage. The right main landing gear strut is in the foreground of the upper part of the photo; the top of the left main landing gear strut on the other side of the aircraft can be seen through the lower access area. (Ben Peck)

Located in front of the two-piece linkage inside of the wheel well is a bar in which the linkage pivots on as seen on this left main landing gear section. (Ben Peck)

The main landing gear section of the TF-102A is identical to the F-102A aircraft. In this view, the scissors assembly is slightly retracted. (David Kidman)

The main landing gear tires are larger than the nose landing gear tire. The main landing gear wheel assembly is also larger and follows a much different design. (Ken Neubeck)

The Delta Dagger aircraft has a very unusual looking landing gear door structure that is a result of the layout of the main landing gear on the aircraft. The lower section of the structure has two red hooks that connect with the fuselage when the door is closed during flight. (Ken Neubeck)

The door panel attached to the left main landing gear assembly hides most of the landing gear from view in this outboard shot. The shape of the door panel conforms with the contours of the wing and the underside of the fuselage into which it merges after takeoff. The two red hooks located towards the outside of the lower door panel engage with the fuselage when the door closes. (Ken Neubeck)

A yellow and black emergency button is located on the left side of the aircraft where the wing meets the fuselage. The button opens a square door behind which is the T handle used to jettison the canopy in case of a ground emergency. (Ken Neubeck)

Behind the armament doors on the left side of the aircraft is an access door where the ground crew can refuel the aircraft. (Ken Neubeck)

The two bomb bay door assemblies meet on the lower fuselage. An armament gear deploys from the aircraft to release the missiles. (Ken Neubeck)

The top warning label warns the ground maintenance personnel of the fast moving doors on the armament bay and the lower label warns about the armament gear. (Ken Neubeck)

The open upper and lower door panels on the left side of an F-102A reveal an empty weapons bay. A boarding ladder appears in the foreground. (Ken Katz)

Located inside the missile bay area is a series of linkages, actuators, and other hydraulic components that are used to open and close the door panels. (Ben Peck)

Two AIM-4 Falcon missiles are attached to extension links on the right side missile bay. AIM-26 and AIM-4 missiles are also extended from the missile bay in the center of the lower fuselage. (Ken Katz)

An F-102A aircraft is on display along with some of the missiles it carries. The white missile is the AIM-26 Super Falcon while the smaller orange missiles are AIM-4 Falcon missiles. (National Archives via Dennis R. Jenkins)

An external pylon may be attached underneath the wing outboard from the landing gear, to carry external fuel cells. (Ken Neubeck)

For extended range, an external pylon and external fuel cell are attached to both of the wings of this F-102A. The tanks would be dropped before any combat. (Ken Neubeck)

There are two external attachments on the outboard side of the pylon for connecting to the external fuel tank, a service upgrade introduced in 1958. (Ken Neubeck)

The 230-gallon external fuel tanks used on the F-102A have a refuel cap towards the front of the tank, and have two inboard attachments on the pylon. (Ken Neubeck)

Two fence structures, each over a yard long, are mounted toward the outboard side of the Delta Dagger's wings. (Ken Neubeck)

The inboard fence structure is located on the top of the wing and extends aft from the wing's leading edge. (Ken Neubeck)

Unlike the inboard wing fence, the outboard fence wraps around the leading edge of the wing and extends back about six inches on the underside of the wing. (Ken Neubeck)

On later-production F-102As, redesigned, downward-pointing wing tips have replaced the earlier, upward-pointed wing tips. (Ken Neubeck)

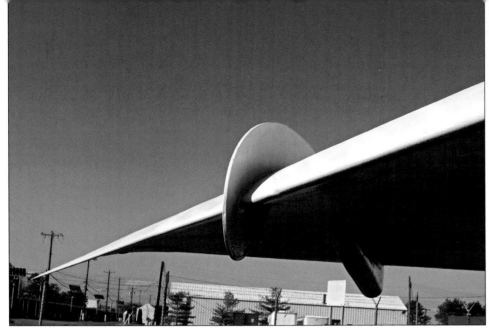

The outboard fence structure wraps underneath the wing. (Ken Neubeck)

The U.S. Air Force national insignia is located on the underside of the wing, inboard of the fence and the fairing structures. (Ken Neubeck)

Located directly behind the outboard fence is a fairing structure that is located on the rear of the wing but is not attached to the elevon surface. (Ken Neubeck)

The fairing under the left wing is identical to that under the right wing. (Ken Neubeck)

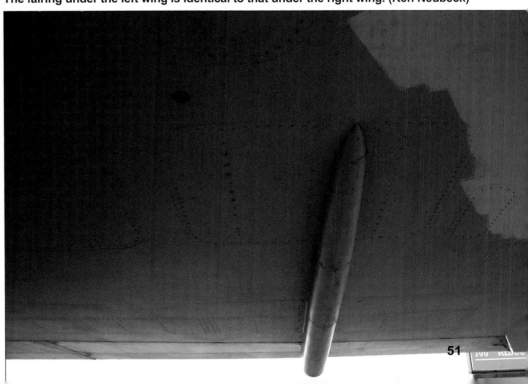

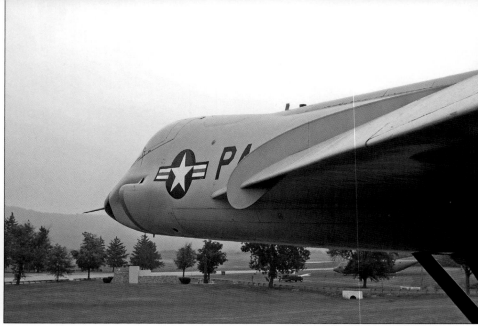

Although the front section of the fuselage of the TF-102A differed significantly from that of the F-102A, the wing design remained virtually unchanged. (Ken Neubeck)

The front fuselage area of a TF-102A appears beyond the aircraft's wing and wing fence. The TF-102A is four feet shorter than the F-102A. (Ken Neubeck)

All TF-102A aircraft featured the Case XX wing configuration that was introduced in later-production Delta Daggers, including the later F-102As and the TF-102As. (David Kidman)

A minor change to the outboard fence on the wing was the addition of a reinforcement plate in the area where the fence went around the leading edge of the wing. (David Kidman)

Four F-102A aircraft in their original light gray paint scheme sit on the tarmac at Tân Sơn Nhứt Air Base in Vietnam in 1964. Base defense was the primary function of F-102As during the Vietnam War, but later the aircraft became involved in ground attacks on enemy buildings and other targets. The pilot boarding ladders are curved to go over the left engine cooling inlet. During Vietnam service, one F-102A was lost in air-to-air combat with a MiG-21 while three others were shot down by anti-aircraft or small arms fire. Additional aircraft were destroyed on the ground during Việt Cộng mortar attacks, or lost due to accidents. (National Archives via Dennis R. Jenkins)

A green position light and a portion of the elevon are seen in this view of the underside of the right wing. (Ken Neubeck)

On the underside of the left wing, there is a red position light that is situated in the middle of the wing tip. (Ken Neubeck)

The elevon on the F-102A aircraft is the single flight-control surface on the trailing edge of the wing between the wing tip and fuselage. (Ken Neubeck)

The elevon and red position light are seen on the upper surface of the aircraft's left wing. (Ken Neubeck)

The elevon on this F-102A is slanted downward. Several hinge points connect the elevon to the wing. (Ken Neubeck)

The hinge point connecting the elevon to the wing consists of a pushrod assembly that fits inside a bracket structure. (Ken Neubeck)

Although most F-102A access panels are screwed in place, there are some quick-release panels that provide access to critical areas in the wing. (Ken Neubeck)

Throughout the lower part of the wings are points for grounding as well as drainage ports. Panels are secured to the airframe by flush Phillips-head screws. (Ken Neubeck)

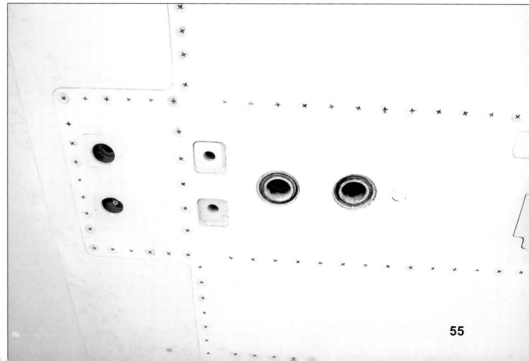

This TF-102A on display at Edwards AFB test center in California features a complete set of warning markings and unit insignias. (David Kidman)

The Air Force Flight Test Center emblem associated with Edwards AFB is emblazoned on the front fuselage of this TF-102A Aircraft. (David Kidman)

The left side of this TF-102A shows the advanced aerodynamic structural design that was specially contoured in the areas around the inlet ducts for the engine – a design based on lessons learned from the F-102A. (David Kidman)

Located above the engine cooling inlet duct is the ejection seat warning marking. A sheet-metal panel has been added on the duct in order to facilitate better airflow. (David Kidman)

A Tactical Air Navigation ("TACAN") blade antenna is located directly in front of the closed nose landing gear door on the underside of this TF-102A aircraft. The access panel on the right forward fuselage section has been contoured with an indentation in front of the engine intake vents. The latch assembly on the access door has been contoured as well. (Ken Neubeck)

Located in the middle of the lower fuselage of the F-102A are two access doors that are hinged on the inside. These doors provide access to the lower portion for the middle section of the engine where there are various lines and maintenance points to access. There are several drainage holes and exit ports in this area of the fuselage as well, along with the lower red landing light assembly. (Ken Neubeck)

The rear of the TF-102A is identical to the rear section of the original F-102A aircraft, having an identical layout of structures as well as similar dimensions. (David Kidman)

A devil motif adorns the tail of this F-102A on display at the McClellan AFB museum near Sacramento, California. (André Jans)

This F-102A aircraft, serial number 56-1247, has served in many locations. It is seen here on display at Travis AFB in California with the tail markings of the 82nd FIS based in Germany. (Ken Neubeck)

TF-102A, serial number 54-1366, is painted in the colors of the 526th FIS that was based in Ramstein Air Base, near Kaiserslautern, in the German state of Rheinland-Pfalz. (Ken Neubeck)

The location of the pitot static tube and the high tail configuration are distinctive tail design features on this F-102A aircraft, serial number 56-1393, on display at the Pima Air and Space Museum, located in Tucson, Arizona. This F-102A is decked out in the markings of the 327th FIS that was based out of George AFB, California. (Ken Neubeck)

During F-102A production, changes were made in the design to increase the height of the tail. Aircraft with the tall tail design can be identified by the squared-off top of the tail section, in contrast to the rounded tip of earlier models. (Ken Neubeck)

In addition to the pitot static tube that is on the nose of the aircraft, there is a set of ram air intakes for providing the feel force for the rudder elevator control system. (Ken Neubeck)

Because of the delta wing design, the F-102A has an unusual aft section that was designed to deal with the area rule of aerodynamics. (Ken Neubeck)

The two stainless steel wasp-handle assemblies were added for better aerodynamic capabilities and are flared with the rest of the fuselage. (Ken Neubeck)

Located inside the two halves of the speed brake assembly is where the drogue chute is located and from which it deploys when the aircraft is landing. (Ben Peck)

The parachute on this aircraft is an under-sized display model. On an operational F-102A, the drogue would be larger and its lines would extend an additional 20 feet. (Ben Peck)

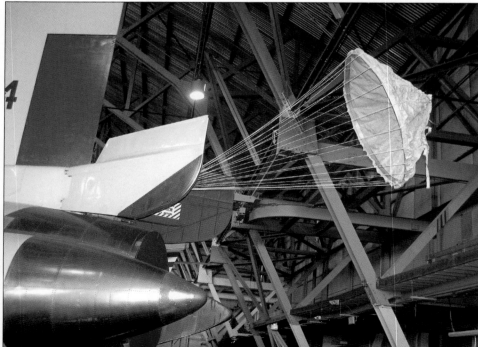

In its unpainted state, this aircraft reveals that two different types of metal were used in the construction of the aft section. (National Archives via Dennis R. Jenkins)

Two bulges in the fuselage and the (closed) speed brake located above them are apparent in this view of the aircraft's rear section. (National Archives via Dennis R. Jenkins)

This early-production F-102A has the original, upward-twisting "Case X" wing tip that was later changed in production. (National Archives via Dennis R. Jenkins)

The speed brakes are open in this view of the rear of a Delta Dagger. (National Archives via Dennis R. Jenkins)

The tail markings for this F-102A aircraft, serial number 57-088, belong to the New York ANG unit based out of Westhampton air base in Long Island. The markings include an abstract bird painting inside of the blue section of the tail. F-102A aircraft were only based at this location for a few years: from 1972 to 1975. (Ken Neubeck)

The tail of this F-102A, serial number 56-1264, features the markings of the Connecticut ANG unit whose Bradley Airport base is where the plane is currently on display. The emblem is that of the ANG unit featuring an American Revolutionary Minuteman. This ANG unit flew F-102A aircraft from 1966 to 1971. (Ken Neubeck)

TF-102A, serial number 54-1353, on display at the Air Force Flight Test Center Museum at Edwards AFB, California, wears USAF test aircraft markings and emblems. The inscription on the coat of arms inside the black slash reads: *"Ad Inexplorata,"* which means "into the unknown" in Latin. The two position lights above the rear portion of the wing are intact. (David Kidman)

This TF-102A, serial number 56-2346, is on display at the U.S. Army post Fort Indiantown Gap in Pennsylvania. Although this particular aircraft was not specifically based there, it served with a Pennsylvania ANG unit during its active service. Tail markings include the emblem of an American Revolutionary Minuteman that is emblazoned on ANG aircraft. (Ken Neubeck)

The vertical tail section of the F-102A, to which the aircraft's airbrakes are attached, is flanked on both sides by wasp handles with stainless steel extensions that are unpainted. Some F-102A exhibit models may have these extensions painted for corrosion protection. The vertical section of the F-102A tail is quite high in comparison to other high-speed aircraft of its time. (Ben Peck)

This close-up view of the top of the tail section shows that two panels riveted together make up the tail assembly. The panels flare together as they proceed upward toward the top of the tail. Just below the tip of the tail is a horizontal structure that provides additional stability to the aircraft in general, and to the long tail section specifically, when the aircraft travels at very high speeds. (Ken Neubeck)

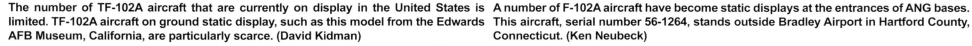

The Convair F-106 Delta Dart replaced the F-102, entering service in the 1960s, and remaining in use through the 1980s. In all, 340 Delta Darts were built. (Ken Neubeck)

This silver-painted F-102A, serial number 56-1416, carries the distinctive markings of the 57th FIS at Keflavik in Iceland where it served during the Cold War.

The number of TF-102A aircraft that are currently on display in the United States is limited. TF-102A aircraft on ground static display, such as this model from the Edwards AFB Museum, California, are particularly scarce. (David Kidman)

A number of F-102A aircraft have become static displays at the entrances of ANG bases. This aircraft, serial number 56-1264, stands outside Bradley Airport in Hartford County, Connecticut. (Ken Neubeck)

The end of the line for almost 200 F-102A aircraft came were when they were converted to unmanned target drones (designated PQM-102A) during Pave Deuce. (National Archives via Dennis R. Jenkins)

By the 1970s, several F-102A aircraft would be decommissioned from the ANG units and end up in reclamation yards. This former Connecticut ANG aircraft was heavily damaged during a tornado that hit Connecticut in October of 1979. (Ken Neubeck)

The F-102A gate guard at Westhampton, New York, ANG was moved inside the base in 2009, making it very limited to public access, except by appointment, a situation that is common for F-102A exhibits in the United States. (Ken Neubeck)

Several museums are restoring F-102A aircraft. This particular aircraft is on loan from the National Museum of the USAF in Ohio to the Empire State Aerosciences Museum in Scotia, New York, for restoration. (Ken Neubeck)

Four straight-tip wing design F-102A aircraft fly in formation on a training mission over rural Texas during the late 1960s. The unique red and blue arrow design painted over a white band on the tail is the symbol of the 4780th ADW that was based at Perrin AFB in Denison, Texas. The 4780th ADW had the largest F-102A fleet from 1962 until May of 1971, when the base closed and the F-102A was removed from regular USAF service. (National Archives via Dennis R. Jenkins)

The F-102A has an arresting hook assembly that deploys from the lower fuselage near the rear of the aircraft for landings on short runways. The arm and hook assembly is seen here in the stowed position on the aircraft. (Ken Neubeck)

This view from the rear of the arresting hook shows the hook assembly in the stowed position on the underside of the aircraft. (Ken Neubeck)

A protective bracket partly covers the hinge assembly where the arm of the arresting hook meets the lower fuselage. The arresting hook release is an all-mechanical system that would be deployed in certain aircraft emergency situations on runways equipped with arresting cables. (Ken Neubeck)

The open left, right, and center weapon bay doors are visible in this shot looking forward from the rear of the underside section of the F-102A. The engine warning stripe extends to the lower fuselage as well, to alert ground maintenance. (Ben Peck)

A protective plate covers the bottom of the red lower landing light on the underside of the fuselage. (Ken Neubeck)

The lower landing light on the underside of the aircraft is seen from the side. The right main landing gear assembly appears in the background. (Ken Neubeck)

In addition to the wing position lights, there are several other position lights located on the F-102A. There are two position lights on the front of each wasp waist. (Ken Neubeck)

There is a position light above the insignia, and below the upper electronics access doors on both sides of the fuselage of the TF-102A. (Ken Neubeck)

There is a red position light located on the lower fuselage, just behind the main landing gear. The landing gears are stowed in this photo. (Ken Neubeck)

Small, circular position lights are fitted flush with the surface of the fuselage just below the leading edge of the wing on both the left and right sides of the aircraft. (Ken Neubeck)

This overhead view of the F-102A shows that the wings are angled toward the rear and that the fuselage slims down at mid-length to comply with aerodynamic forces. (USAF)

A pilotless PQM-102 (s/n 56-1254) flies over Tyndall AFB in Florida in 1980. Many F-102 aircraft were converted to pilotless drones, beginning in 1974. (USAF)

F-102A aircraft (s/n 56-1260) in Bitburg, in the West German state of Rheinland-Pfalz, was part of the 525th FIS Bulldogs during the mid-1960s. (USAF)

This TF-102A (s/n 54-1366), painted as one of the 525th FIS Bulldogs, is on display at the Pima Air and Space museum in Tucson, Arizona. (Ken Neubeck)

Located on the lower fuselage is a TACAN AN/ARN-21 blade antenna that is mounted on a plate directly in front of the nose landing gear wheel well, which is closed on this TF-102A. (Ken Neubeck)

Also located on the lower fuselage, in front of the TACAN blade antenna and just behind the nose cone, is the circular-shaped IFF antenna that is held in by a series of Phillips screws. (Ken Neubeck)

The UHF Command AN/ARC-34 radio antenna is located on the spine, directly behind the cockpit. A red position light is located to the rear of this antenna. (Ken Neubeck)

The trapezoidal-shaped structure located on the top of the horizontal tail section is the AN/ARN21 VHF antenna localizer. (Ken Neubeck)

Attached to the ceilings of both the weapon bays are dual sets of large brackets. The brackets in the right weapon bay are seen here from the rear of the bay. Each set of brackets is attached to a rail assembly on which each air-to-air missile is mounted. The bracket assemblies are hydraulically driven and deployed in flight when the bay doors are opened. (Ben Peck)

Directly behind the right weapon bay and before the right main landing gear is an access door that houses the Ram Air Turbine (RAT) for emergency hydraulic power. If the aircraft loses hydraulic power during flight, the pilot commands this door to open, allowing the turbine to deploy and rotate, driving the hydraulic system. Hydraulic lines connect the RAT to a hydraulic reservoir located in the same bay. (Ben Peck)

This view from the left side of the lower fuselage shows armament bay access doors closed, along with an extruded assembly in the middle of the fuselage. (Ken Neubeck)

The extruded assembly on the lower fuselage is a housing that contains a release handle for the armament bay access doors. (Ken Neubeck)

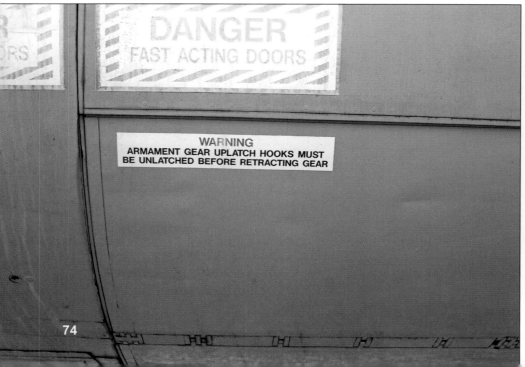

Because of the inherent danger in the release of the three sets of armament doors on the ground, there are warning labels affixed to all of the doors. (Ken Neubeck)

Ground maintenance personnel use this release handle, seen here up close, to open the armament bay access doors on the underside of the aircraft. (Ken Neubeck)

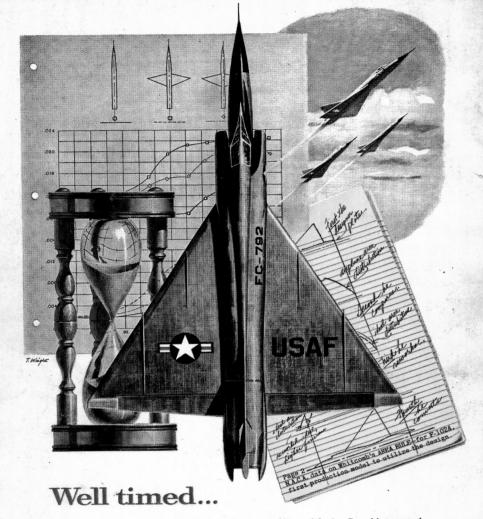

During the 1950s, the Convair Company would place many advertisements for its aircraft products, including the F-102A Delta Dagger, in aerospace periodicals and even general magazines. These ads provide significant details of the aircraft that can be seen clearly, more clearly even than in photographs. In the ad shown on the left, for example, the canopy area is painted in exceptional detail, especially the top portion of the ejection seat, showing the red headrest and the top of the seat, which was capable of breaking through the canopy during ejection. In the ad on the right, the top of the F-102A is painted in detail. The four fences on the wings, the shape of the spine, the canopy shape, and the elevon surfaces show up in excellent detail. (Convair Archives)

Bright colors were a common feature on many of the F-102A aircraft that were assigned to colder, snowy climates. The red tail section of this aircraft from the 317th FIS based out of Elmendorf, Alaska, would facilitate visual location and rescue in snow.

The color red is also prominent on the aft section of this F-102A from the 59th FIS, which was based in snowy Keflavik, Iceland.

This aircraft sports the markings of the 4780th ADW, based out of Perrin AFB in Denison, Texas, home of the primary flight interceptor training school.

During the early stages of the Indochina War, F-102A aircraft retained their original gray paint scheme when deployed to Tân Sơn Nhứt Air Base in Vietnam.

Later during the Vietnam War, F-102A aircraft were painted in the Southeast Asia (SEA) camouflage paint scheme.

F-102A aircraft also deployed to West Germany, where they served with the 526th FIS based out of Ramstein AFB.

This aircraft serves the 111th FIS, a part of the Texas Air National Guard (ANG). Beginning in 1960, F-102A aircraft were assigned to many ANG units.

This aircraft wears the markings of the 196th FIS, which was part of the California ANG when the squadron converted from the F-86D to the F-102A in 1965.

Beginning in 1978, the Sperry-Rand Corporation modified several F-102A aircraft into pilotless drone planes with the designation QF-102A.

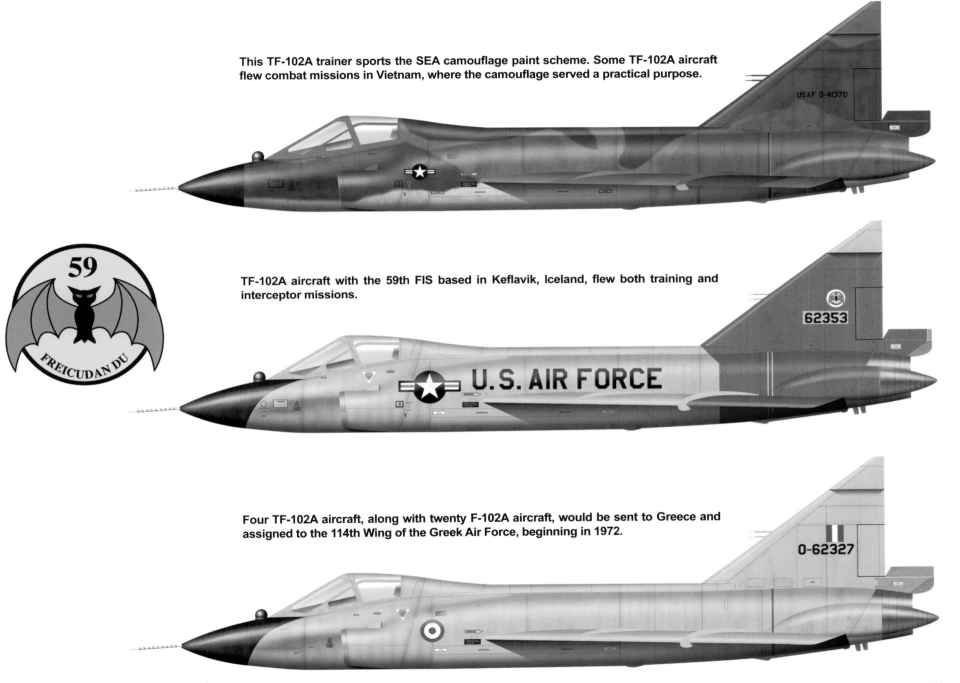

This TF-102A trainer sports the SEA camouflage paint scheme. Some TF-102A aircraft flew combat missions in Vietnam, where the camouflage served a practical purpose.

TF-102A aircraft with the 59th FIS based in Keflavik, Iceland, flew both training and interceptor missions.

Four TF-102A aircraft, along with twenty F-102A aircraft, would be sent to Greece and assigned to the 114th Wing of the Greek Air Force, beginning in 1972.

Two F-102A Delta Daggers assigned to the 57th FIS depart from a U.S. runway. The F-102A was primarily successful in the interceptor role for the USAF in stalking Soviet aircraft during the Cold War. There was little need for a supersonic interceptor in Vietnam, so its use there was limited. Eventually the F-102A was replaced by the F-106 Delta Dart. (National Archives via Dennis R. Jenkins)